//:the_
flickering_
page/

ATELIER26 : samizdat series, Vol.2 : 2014

PORTLAND, Oregon

the_flickering_ page/

the reading experience
in digital times

3 guided chapters (and a preface) by
M. ALLEN CUNNINGHAM

Illustrated by
NATHAN SHIELDS

Illustrations and cover design by Nathan Shields
Book design by M.A.C.

isbn-13: 978-0-9893023-2-6
isbn-10: 0989302326

Library of Congress Control Number: 2014904584

This book is adapted from a presentation given by M. Allen Cunningham
as part of the Oregon Humanities Conversation Project, 2011-present.
Sales proceeds do not benefit Oregon Humanities.

Atelier26 Books are printed in the U.S.A. on acid-free paper

"A magnificent enthusiasm, which feels as if it never could do enough to reach the fullness of its ideal; an unselfishness of sacrifice, which would rather cast fruitless labour before the altar than stand idle in the market."—John Ruskin

Atelier26

Atelier26Books.com

//:contents/

//:a_preface/

5_characteristics_of_an_"e-book world"/

Let's start with a short list of observations about our widespread adoption of e-reading technologies, what this behavior says about us, and where it may be leading us.

One could say ...

1) We like books that are procured without trouble, that are quickly read, that deliver *themselves* and that we can enjoy right away.

2) Authors will aim for *faster* rather than *finer* writing. Increasingly, speed of production will trump quality.

3) Short-form writing will become more common than big, long books.

4) Liveliness will become more common than learnedness; fantasy and virtual reality more common than depth and feeling.

5) Writers will try to surprise, startle, or upset us, to grab our attention by gimmicks or provocation rather than by the subtleties of their art.

Such comments obviously represent a pessimistic view — and in some contemporary corners such things have been said about e-books. All of the above observations, however, date from more than a century and a half ago. I've paraphrased them from Part Two of Alexis de Tocqueville's great book *Democracy in America,* published in 1840. Tocqueville was commenting on the state of the American book market — and the predilections of American readers — as he found them at that time, and he was envisioning how literary culture might develop within the democratic ethos of the New World. Here are Tocqueville's actual words:

"Having only a short time to give to letters ... they like books that are procured without trouble. ... They demand facile beauties that deliver themselves and that one can enjoy at that instant. ...

Authors will aim more at rapidity of execution than at perfection of details. Small writings will be more frequent than large books, spirit than erudition, imagination than profundity. ... One will try to astonish rather than to please, and one will strive to carry away passions more than to charm taste."

A major allure of digital publishing, equally celebrated and decried, is the way this new technology allows anyone to circumvent the "gate-keepers" of traditional publishing. Writer or not, if you wish to produce and distribute a text you are immediately empowered to do so by use of the Internet, no editorial approval required. Let's hear from another commentator made squeamish by this newfound freedom:

> "Now that anyone is free to print whatever they wish, they often disregard that which is best and instead write, merely for the sake of entertainment, what would best be forgotten. ... And even when they write something worthwhile they twist it and corrupt it to the point where it would be much better to do without such books, rather than ... spreading falsehoods over the whole world."

As remarkably contemporary as it seems, this commentary comes to us from 1471, penned by the Italian scholar Nicolo Perotti in the immediate aftermath of Gutenberg's printing press, developed just sixteen years earlier. Perotti's alarm is a response to the tsunami of new books and would-be authors scrambling to publish.

to_sum_up/

Historical perspective is helpful. While discussing the implications of digital publishing and e-reading, let us bear in mind that our particular moment of cultural transition (or upheaval, as many would call it) is not the first of its kind.

//:chapter_1/

The Technology

of

Individualism©

How Print Helped Shape the Renaissance

On the preceding title page we find illustrated:

- Moveable type
- Michelangelo's "David"
- A printed book
- Fragonard's "Reader"
- Martin Luther
- Thomas Jefferson
- Ralph Waldo Emerson
- The FBI Warning (and copyright symbol)

This chapter will consider how these seemingly random figures and things correspond under the phrase "the technology of individualism."

a_prophet_misperceived/

Our guide for this chapter is **Marshall McLuhan**:

McLuhan's fame as a public intellectual began with the appearance of his 1964 book *Understanding Media: The Extensions of Man,* in the pages of which he deployed his most famous — and most misunderstood — phrase:

"The Medium is the Message."

In one of the great ironies of intellectual history, "the medium is the message" was expropriated by media-makers themselves for use as the nihilistic credo of their trade. As we all know too well, the fixing of the public's attention upon a 24-hour marketing spectacle became, for most mass media in the second half of the twentieth century, an end in itself. Toward this end, McLuhan's words were widely invoked as meaning "so long as a medium is all-powerful, no message is necessary."

In reality, "the medium is the message" was a cautionary call. McLuhan was urging a better understanding of how each technological tool, each medium we employ influences our behavior, our social norms, our very consciousness. As he put it,

> **"A medium molds what we see and how we see it. Eventually, if we use it enough, it changes who we are as individuals and as a society."**

Let's turn now to one of this man's lesser known quotes:

> **"Print is the technology of individualism. If man decided to modify this visual technology by an electric technology, individualism will also be modified."**

These astonishing words appear in McLuhan's 1962 book *The Gutenberg Galaxy: the Making of Typographic Man.* In the first primitive days of the so-called computer age, McLuhan forecasted the cyber-social culture that currently envelops us all, a culture characterized by a terminology we hardly think about, e.g., "social media," "connectivity," "interactivity," "status update," "Facebook friend." *Modify print technology with electricity, and individualism will also be modified.*[1]

Considering McLuhan's preternatural insight — we may as well call it *prophecy* — we should note that he's scrupulous in his terms. He does not say anything as general as, "Books are the technology of individualism," but refers specifically to the primary technology by which books have been reproduced en

---

[1] In a recent book entitled *Elsewhere U.S.A.*, author Dalton Conley coins a McLuhanesque term for a new kind of person incubated in the always-connected electric technology of social media: the *Intravidual.*

masse since the mid-fifteenth century, the technology of actual pages bearing actual ink: Print.

But what makes print the "technology of individualism"? To best get McLuhan's gist, let's look back at the history of the book and how the printing press revolutionized culture.

Here we have the man credited with developing the technology of individualism:

Johannes Gutenberg

Johannes Gutenberg perfected moveable type and the processes of the printing press. These were processes long existent in other cultures, but Gutenberg made several key advancements, and around the year 1455 he printed approximately 180 Latin copies of the "Gutenberg Bible" in Germany.

At this point, says McLuhan, "Europe moved into the age of the machine," a statement slightly counterintuitive. *Books* gave birth to the machine age? Since the avid readers among us tend to approach the object of the book almost entirely from a humanities perspective, the notion that bookmaking was integral to the history of *industrial* development can take some getting used to. But so it was, for with that first machine-made book,[2] the world was introduced to mass production.

---

[2] Machine-made though it was, the Gutenberg Bible was created to resemble as closely as possible a scribal manuscript: double columns, dropped capitals, and embellishments and illuminations added in color by hand after printing. For a long time following Gutenberg, a print job's quality was judged by how closely it approximated work done by hand.

Before the printing press, never in the western world could innumerable identical copies of a commodity be laid before a paying public for consumption. As McLuhan notes, by making possible a mass-produced product, the printing press "created the public" as we understand that term.

From this foundation of new technology and new social reality would grow some powerful cultural values that we still cherish today. In their time of emergence, these values helped shape the Renaissance.

As Alberto Manguel notes in *A History of Reading* (Penguin, 1996), "Often a technological development — such as Gutenberg's — promotes rather than eliminates that which it is supposed to supersede, making us aware of old-fashioned virtues we might otherwise have overlooked or dismissed as of negligible importance."

24

a_culture_revolutionized/

To fully appreciate the profound cultural impact of print technology, we must bear in mind that before Gutenberg, books were …

- **Hand-copied** by scribes.
- **Restricted** to private libraries or owned by the wealthy elite (a precious object, a book would often literally be chained to a shelf or monastery lectern).
- **Often inconsistent** in text from one copy to another (since every copy was a one-of-a-kind object, there was no such thing as an "authoritative text" as we understand that term today).
- **Not subject to copyright** (the concepts of plagiarism or infringement were nonexistent, let alone the notion of "owning" sentences).

Pre-Gutenberg, books were not the business of the common people. A book's content was more likely to be heard than read. Usually illiterate, the commoner received information, stories, poems, etc., from an oral source in the public square or in church. It was in church particularly, where religious texts were doubly filtered through inscrutable Latin and the doctrinal biases of clergical authorities, that orality often disempowered the commoner. Print changed this.

Relative to the pace of historical change, the cultural transformation wrought by Gutenberg's press was sudden and dramatic. In quantity alone, print books were revolutionary: during a time when the collections of the greatest university libraries rarely surpassed 100 titles each, Gutenberg's first print run of Bibles itself was nearly twice that large. Historian Henry Petroksi notes that by the end of the sixteenth century in Europe, more than 100,000 different printed titles were circulating. "If it is conservatively assumed that there were on average as few as one hundred copies of each book (print runs of several hundred were not uncommon in the fifteenth century), ten million individual copies of books were available to Europeans."[3]

Print not only made books widespread beyond the walls of private libraries — it made books more generally portable. No longer massive one-of-a-kind objects chained to monastery lecterns, volumes were often pocketsize now. Smaller in format and newly abundant, books became an affordable expense for readers outside of the noble class. Amid the sea of printed titles, literacy spread widely, and as literacy grew, further books were printed and sold, and before long literature in its freshly portable form became something other than a strictly collective experience. Reading in private became commonplace.

Books had moved into the realm of the common man. *Print technology created the public.*

---

[3] *The Book on the Bookshelf* by Henry Petroski (Knopf, 1999)

say_hello_to_modern_man/

With the mass consumption of books by a paying, literate public, several new cultural values grew to carry considerable importance. For instance, there developed a special emphasis on the author as a figure of cultural authority, and from this evolved the concept of discrete disciplines and schools of thought. As historian James Burke describes, "By ending the copying errors with which manuscripts were rife, [print technology] placed on the author the responsibility for accuracy and definitive statement. ... This led to the concept of 'mastership' in a subject, which in turn led to the fragmentation of knowledge into specialized areas, emphasizing the separation of the 'expert' from the rest of the community."[4] Now you really needed to "know your stuff" if you intended to publish your learning for the world to read; somebody else may publish a

---

[4] *Connections* by James Burke (Little, Brown, 1978)

competing claim to the public's attention, after all. Thus **objectivity** became a key intellectual practice and driver of the sciences.[5]

As print technology steadily drew people away from the collective traditions, artistic and intellectual conventions, and centralized authority advanced by oral culture, another quality would come to be newly enshrined alongside objectivity, and this too we continue to prize today: **subjectivity**. That is, print allowed for and encouraged the marvelous expression of individual perspective, personal genius, and artistic idiosyncrasy. Author Nicholas Carr, another guide we will look to shortly, helpfully conceptualizes all the rich transformation of this post-Gutenberg period as the result of "a new intellectual ethic: the ethic of the book. ... [After] quiet, solitary research

---

[5] Another eventual offshoot of the new focus on authorship would be the development of the notion that a published idea is a commodity, a thing one can own and potentially profit from. Copyright would be born.

became a prerequisite for intellectual achievement, originality of thought and creativity of expression became the hallmarks of the modern mind." Historian Kenneth Clark refers to printing as "the extension of [man's] mind through the word," which we may understand as meaning innumerable, singular minds extending to innumerable other single minds in a way never before imagined in western culture.

Or, as McLuhan puts it:

> **"Until the modern movie, there had been no means of broadcasting a private image to equal the printed book."**

"The reader of print," McLuhan elaborates, "stands in an utterly different relation to the writer from the reader of manuscript. Print gradually made reading aloud pointless, and accelerated the act of reading till the reader could feel 'in the hands of' his author."

Newly private and newly personal in all of its aspects, the kind of reading facilitated by print powerfully fueled **individualism** as a cultural value. Meanwhile, the shift away from church authority enabled a resurgence of classical (i.e., "pagan") art and philosophy through the printing of ancient texts,[6] and thus **humanism** found a new social prominence.

---

[6] The fifteenth century Venetian Aldus Manutius pioneered many aspects of the art of printing, and attained preeminence as the publisher of accurately reproduced classical texts which, before his time, were error-laden and extremely hard to come by. "These [books], which for a long time were only to be found here and there, may go into the hands of scholars in a more correct form," he wrote in the 1495 preface to his edition of Cicero's *The Art of Rhetoric*, one of Aldus's first publications.

Objectivity, subjectivity, individualism, humanism: all were propelled by print, and all were integral components of that astonishingly fertile and ingenious age we call the Renaissance.

Leonardo DaVinci's Vitruvian Man, created in 1490, expresses vividly and concisely the new world-view of the Renaissance.

DaVinci developed the figure on the basis of an architectural theory by the first-century Roman engineer Vitruvius. According to Vitruvius, the harmonious measurements within an ideal human form — each section of the body precisely and proportionally related to every other — would, when applied to building, produce ideally beautiful architecture. DaVinci's successful visual interpretation of the Vitruvian theory is pure post-Gutenberg humanism. As historian Thomas Cahill puts it, here is "a vision that makes man the measure of all things ... a world made (or remade) to a human scale."

How far we have come from the God-fearing, collective, authority-driven traditions of an oral Europe. Print was a primary factor in this sea change.

to_sum_up:_renaissance_values/

After the birth of print in Europe, objectivity became a critical aspect of intellectual inquiry as specialized disciplines took hold. Subjectivity, fostered through the idiosyncratic expression and perspectives found in books, grew to be valued as an intellectual and aesthetic principle. Humanism, bred of revived classical art and texts, shifted the spotlight from the authority of an aloof omnipotent Creator to the wonder and glory of his creations. And individualism took root in a newfound access to printed works and the freedom and ability to experience books – including, eventually, the Bible – in solitude and without intermediary.

To return to our title illustration, we may now see more clearly the relation between its emblematic figures:

The Technology of Individualism

- **Moveable Type:** Small in itself, its socio-cultural impact was immeasurable.
- **Michelangelo's "David":** Perhaps more dramatically than any other, Michelangelo exemplifies the Renaissance ideal of the artist as hero, whose work celebrates the glory of man and woman as God's creations. Though he was devoutly Christian, Michelangelo's visual emphasis is always on the humanity of his figures – even where the figure in question is God himself. Born in 1475, a mere twenty years after the Gutenberg Bible, Michelangelo was a child of print.
- **Fragonard's "Reader":** The very image of readerly privacy. Alone, she communes deeply with her book (a novel, perhaps?).
- **Martin Luther:** In 1517 he kickstarts the Protestant Reformation, a violent move away from church authority and toward direct personal access to God's word. In 1521 Luther publishes a German Bible. He

refers to the printing press as a gift from God to the people.

- **Thomas Jefferson:** A great Enlightenment mind, individualist, and advocate for personal liberty and the pursuit of happiness.
- **Ralph Waldo Emerson:** Publishes the essay "Self Reliance" (1841), a gospel of American individualism. In all his writings, Emerson advocates nonconformity and celebrates personal destiny and the preeminence of individual conscience.
- **FBI Warning/Copyright Symbol:** A direct legacy of print technology, the ownership of "intellectual property" is an idea very much with us today.

Each of the cultural figures above was a direct beneficiary of print, the "technology of individualism." Each is a benefactor to all of us, of individualism as a cultural force.

//: read_me/

From *Hamlet's Blackberry: Building a Good Life in the Digital Age* by William Powers (Harper, 2010)

"Lately there's been an effort to make reading, the ultimate inward experience, more outward. Some e-reading devices allow you to toggle your attention back and forth between the text and the rest of the digital universe — the always-connected book. Enthusiasts of this approach predict that in the future all reading will be done effectively in public. That is, we'll be navigating links, comments, and real-time messages from distant others even as we try to read, say, a terrific novel. In a way, that would be a return to the pre-Gutenberg era, when the crowd looked askance at solitary, silent readers. ...

For research purposes, this Google age is a wonder. But there's a difference between *access* to information and the *experience* of it. Reading evolved away from the crowd for a reason: it wasn't the best way to read. ... Hopping around among competing digital distractions, it's impossible to go truly inward, to become immersed in reading to the point where the crowd falls away, an experience poet William Stafford captured nicely in the lines

> Closing the book, I find I have left my head
> inside.

The point of the new reading technologies, it often seems, is to *avoid* deep immersion, precisely because it's an activity the crowd can't influence or control and thus a violation of the iron rule of digital existence: Never be alone. Deep, private reading and thought have begun to feel subversive. A decade ago, the digital space was heralded for the endless opportunities it offered for individual expression. The question now is how truly individual — as in bold, original, unique — you can be if you never step back from the crowd."

key_questions/

Marshall McLuhan proposed that by asking the following questions of any new technology, we will come to see that tool in a new light. We might ask these questions about e-books and e-reading devices:

1. What does this tool **enhance**?
   (what will I gain in using it?)

2. What does this tool **obsolesce**?
   (what will I lose in using it?)

3. What does this tool **retrieve** from the distant past?

4. How will this tool **turn on me** when taken too far?

further_recommended_reading/

- *The Gutenberg Elegies: The Fate of Reading in an Electronic Age* by Sven Birkerts
- *A History of Reading* by Alberto Manguel (Penguin, 1996)
- *The Late American Novel: Writers on the Future of Books* edited by Jeff Martin and C. Max Magee (Soft Skull, 2011)
- *The Medium is the Massage: An Inventory of Effects* by Marshall McLuhan and Quentin Fiore

//:chapter_2/

Technology

and

Ideology

Why Our Tools Are Never Neutral

the_air_we_breathe/

At first, the relationship (let's call it symbiosis) between technology and ideology can be tricky to grasp. In our wondrous new
age of handheld devices, on-demand viewing, single-track
downloads, and digital commuting, we tend to think of our
technologies as liberating, convenient, and user-defined. We
adjust our tools, we believe, to maximum lifestyle relevance.
Thumbs at the controls, fingers tapping, we summon words
and pictures at light-speed from every corner of the globe.
Choosing what we'll see, we feel choice empower us. With the
press of a few buttons, we tailor-make our work regimens and
customize delivery of our leisure viewing. The Internet itself
we celebrate as the ultimate user-driven technology. We decide
what content to put online, what content to download, and
whether to use or dismiss the incalculable particles of data
encountered en route to the information or entertainment we
desire.

To suggest that such massively useful technologies are in any way ideologically infused can sound, initially, like the stuff of conspiracy theories. However, if one takes the time to look at the history of technological progress, one begins to see that every technological change inevitably carries forward a specific system of ideas or world-view, or promotes a new but no less specific one. Conspiracies are unnecessary — what's at play here is basic human nature. However sleekly robotic or self-enclosed they may seem, our technological products (before they are tools, let's remember, they *are* products) are dreamt up, designed, and utilized by humans, and humans always behave within certain manners and constructs of thought.

Ideology is the air we breathe, and we breathe it into our tools.

a_disclaimer/

In the context of our tumultuous times, we tend to think of ideology as mainly political or religious in nature, e.g., fundamentalist religion or party politics as ideology.

Definitions are useful, especially with as loaded a term as this one. In the present chapter, while considering technology as a vehicle for ideology, we'll employ the latter word in its most general sense, as simply a *system of ideas* or a *manner of thinking.*

Ideology
noun

1 the **system of ideas** at the basis of an economic or political theory **2** the **manner of thinking** characteristic of a class or individual.

picturing_ideology/

Here's a helpful way to visualize the relationship between ideology and technological development. In this illustration,

we see Chartres Cathedral (left). Built over a period of roughly four hundred years in an age we consider to have been extremely "low-tech," Chartres still stands as a stupendous technological achievement. By comparison, consider the neighboring illustration of a metropolitan skyline bristling with skyscrapers. Every such skyscraper is itself a technological marvel. But to juxtapose these two images is to prompt questions about the respective ideologies at work behind technological achievements. We may ask ourselves:

What kind of ideology gives us a cathedral, built in a small village by an anonymous workforce over almost half a millennium?

And:

What kind of ideology gives us a skyscraper, or an urban skyline full of them?

50

The two structure types may share architectural principles, and both stand as evidence of huge technological ambition, and yet each is a distinctive ideological undertaking. Chartres Cathedral and the Chrysler Building, built under different systems of thought, emphasize very different things.

where's_the_neutral?/

Our guide for this chapter is the late cultural critic **Neil Postman**, best known for his 1985 book about the effect of TV on American culture and politics, *Amusing Ourselves to Death: Public Discourse in the Age of Show Business.* In a later book called *Technopoly* (Knopf, 1992), Postman wrote that despite our fond assumption that our tools are merely useful appurtenances, purpose-serving things with no "agenda" in themselves, all technology is in fact inherently ideological. This is because ...

"New technologies ... alter those deeply embedded habits of thought which give to a culture its sense of what the world is like — a sense of what is the natural order of things, of what is reasonable, of what is necessary, of what is inevitable, of what is real. ... A new technology does not add or subtract something. It changes everything."

In other words, new technologies bring on a new *manner of thinking.* Elsewhere in *Technopoly,* Postman was even more incisive:

"Embedded in every tool is an ideological bias, a predisposition to construct the world as one thing rather than another, to value one thing over another, to amplify one sense or skill or attitude more loudly than another."

By appearance, this is a highly provocative comment. But Postman did not mean to suggest that every technological development is inevitably part of some outright campaign to promote a certain ideology. On the contrary, as he pointed out, the ideological bias of a technology is often unforeseeable: the tool's effects can't be predicted, even by those instrumental in its creation. Nevertheless, the link to ideology always exists: technologies are never simply neutral.[7]

---

[7] Postman illustrates this point using the example of the invention of the mechanical clock by Benedictine monks of the twelfth and thirteenth centuries. The clock was originally meant to lend greater regularity to the monastery's canonical hours. But as Postman notes, "By the middle of the fourteenth century, the clock had moved outside the walls of the monastery, and brought a new and precise regularity to the life of the

a_backward_glance/

As we look back through history, we readily find examples of the interplay between ideology and technological development:

---

workman and merchant. … Without the clock, capitalism would have been quite impossible. The paradox, the surprise, and the wonder are that the clock was invented by men who wanted to devote themselves more rigorously to God; it ended as the technology of greatest use to men who wished to devote themselves to the accumulation of money."

54

Ideology	Technology
Manifest Destiny *(continental conquest)*	**Transcontinental Railroad** *(machines triumph over distance, terrain, and indigenous ways of life)*

Ideology   **Technology**

Time as Commodity
("Time is money!")

Standardized Time
*(a synchronized globe
to aid commerce)*

Los Angeles   **New York**   **London**   **Tokyo**

Ideology	Technology
Industrial over Agrarian *(the "ambitious" flee farms and village life for urban centers)*	**Mechanization** *(machines take over human work)*

Ideology	Technology
Mass Production over Craftsmanship	Factory Production
(quantity over quality)	*(machines yield wealth through volume)*

Ideology	Technology
Division of Labor *(a worker is just one part of a machine)*	Assembly Line *(machines demand conformity)*

a_powerful_surround/

Here's more from Neil Postman:

"In the year 1500, fifty years after the printing press was invented, we did not have old Europe plus the printing press. We had a different Europe.

After television, the United States was not America plus television; television gave a new coloration to every political campaign, to every home, to every school, to every church, to every industry...

And that is why the competition among media is so fierce. Surrounding every technology are institutions whose organization — not to mention

their reason for being — reflects the world-view promoted by the technology."

The "world-view" a technology promotes is, naturally, its *ideology.* And since Postman mentions the institutions inevitably surrounding any given technology, we might, for our purposes, call to mind three:

Google

Our instant recognition of the logos above speaks to the massive power and reach of the corporations they represent.

As we all know from our own daily — often hour-to-hour — involvement with these companies, each has a great interest in the public's widespread adoption of its products and, yes, adoption of the *world-view* these products promote. It benefits a corporation when we use the tools it produces. Does it also benefit us as individuals? That question we each must answer for ourselves. But Neil Postman encourages us to ask of any new technology,

> **"To whom will the technology give greater power and freedom? And whose power and freedom will be reduced by it?"**

Contemplating the cultural supremacy of the above corporations, we do well to bear in mind the unequal power dynamics in play: as individuals interrelating daily with these giant firms, we entrust to them an ever-increasing supply of information — about our friendships, our families, our faiths, our political leanings, our buying habits, and our reading experiences — information from which the corporations *always* stand to gain, without respect to our personal privacy, and with no obligation to compensate us beyond our continued use of their tools (and even this is not guaranteed).

infinite_improvements/

Nowadays, we believe almost unquestioningly in "technological progress." That is, we generally accept that there are productive, benevolent, forward-marching processes at work in the world leading us toward:

-   ever-improving functionality, usefulness, and efficiency in our tools *(the clunky desktop computer becomes the portable laptop becomes the pocketsize "smartphone")*

- ever smoother "interfaces" between our tools and ourselves *(dial-up Internet service becomes high-speed wi-fi becomes Google Glass becomes Web implant)*
- ever more just and harmonious global conditions created by virtue of those tools *("Internet 1.0" becomes social media becomes the Arab Spring)*

So widely embraced is this conception of technological progress that it has grown to be one of our reigning ideologies. Author Kevin Kelly's recent book *What Technology Wants* (Viking, 2010) epitomizes this. Kelly, an editor at Silicon Valley's *Wired Magazine,* advances in his book the idea of technology as an evolutionary force, a self-perpetuating juggernaut driven by autonomous "needs" and "desires." As for humankind, we exist mainly to serve and follow technology's primeval forward drive. Another recent book by self-described "futurist" Byron Reese is, in title alone, an advertisement for today's defining ideology: *Infinite Progress: How the Internet and Technology Will End Ignorance, Disease, Poverty, Hunger, and War* (Greenleaf, 2013).

One may best understand the immediate implications of the technological progress ideology by asking oneself: What are the ideology's defining characteristics and how do these permeate my own life? What driving values of this ideology do I and those around me experience on a day-to-day basis?

In answer, the following bywords may well leap to mind:

Speed
Convenience
Portability
Connectivity

These, our chief technological standards, are universally cherished. Our newest tools minister to — and perpetuate — a demand for them all. Each is perceived as a prerogative by almost anyone living amid today's "infinite progress" (a global minority, by the way). Speed, convenience, portability, and connectivity not only drive our technological consumerism, but increasingly color all aspects of our experience, even the heretofore slow, somewhat cumbersome, necessarily solitary act of reading.

disguised_absurdities/

A pair of early advertisements for the Amazon Kindle are typical in the way they encourage a largely unquestioning acceptance of the ideology our gadgets foster. The following are transcripts of these ads:

Ad #1
GUY stands looking at his Kindle. GIRL enters.
GUY: Hey, where you goin'?
GIRL: I wanna get a book that came out today.
GUY: Me too.

GIRL: Come to the bookstore with me.

GUY: I'm good. Got it. *(Shows GIRL Kindle)* It takes less than sixty seconds to download a new book on my Kindle.

GIRL: Sixty seconds, wow! *(Gasps)* That's the book I was gonna get! ...

GIRL stares into the Kindle screen, mesmerized. GUY watches her.

GUY: *(Gloating)* Weren't you going to the bookstore?

GIRL: *(Holds up a shushing finger)* Shh!

GUY smirks knowingly.

Note here the insinuated stupidity of GIRL. Unschooled in the speed and convenience of the technological world, she still considers a trip to the local bookstore to be worthwhile. *Get with the program,* says the new ideology, helpless to tabulate such largely intangible benefits as might come of visiting an actual store; among many other things, these may include:

- the chance discovery of a new book or author
- learning of an upcoming booksigning
- striking up conversation with a fellow customer
- investing in one's community by supporting a local business

The ad's message brings to mind an aphorism by writer Max Frisch: "Technology is the knack of so arranging the world that we don't have to experience it."

Ad #2

GUY stands looking at his Kindle. GIRL enters carry-ing a large shoulder bag.
GUY: That is a giant purse!
GIRL: This can hold two books, two newspapers, three magazines. Pretty great, huh?

GUY: *(Unimpressed)* Yeah. My Kindle holds up to thirty-five hundred books, magazines, newspapers, and it only weighs eight and a half ounces.
GIRL: Yeah, but then I wouldn't get to carry my giant purse.
Pause, as GUY looks at her condescendingly.
GIRL: *(Her purse thuds to the floor)* Can I see that?
GUY: Yeah. *(Hands her the Kindle)*
GIRL: Wow!
GUY: Yeah.

"It is no accident that our nation, the most advanced in technology, is also the most advanced in advertising," wrote our late Librarian of Congress Daniel J. Boorstin in his 1978 work *The Republic of Technology: Reflections on Our Future Community*. "Technology is a way of multiplying the unnecessary. ... Driven by 'needs' for the unnecessary, we remain impotent to conjure the needs away."

The premise of Kindle Ad #2 is that carrying 3,500 books around at all times is more reasonable and sensible than to "lug" a few books in one's handbag. The most obvious question — i.e., How many readers have need of 3,500 separate titles in any given moment? — we are not expected to ask. It's considered beside the point that the technological rationale invoked here (a purely quantitative one) has very little to do with an ordinary reader's experience or preferences. What technological progress has made possible is shown to be, for any "reasonable" person, preferable as well. The intended effect of such ideological messages about technology is, as Boorstin suggests, our unquestioning adoption of certain habits and behaviors which, in a clearer state of mind, we might have deemed absurd. "We live and will live in a world of increasingly involuntary commitments."

72

imposed_changes/

"Books are the last bastion of analog," said Amazon.com CEO Jeff Bezos at the 2007 launch of the Kindle. "Music and video have been digital for a long time, and short-form reading has been digitized, beginning with the early Web. But long-form reading really hasn't. ... [The Kindle] is the most important thing we've ever done." It was well-nigh time, went Bezos's message, for everyone to acknowledge that the printed book (slow to produce, often heavy to carry, lacking buttons or lights) was a medium in need of "an update." Given our wide-spread acceptance of the technological progress ideology, this would seem to make perfect sense. "It's so ambitious," claimed Bezos, "to take something as highly evolved as the book and improve on it. And maybe even change the way people read."[8]

---

[8] It is telling that Bezos should admit his wish to change the way people read (a statement author Sherman Alexie has labeled "imperialist"). As we'll see in the following chapter, e-reading technologies *are* changing

The zealous CEO's words bring to mind a comment from the ancient world:

"An inventor is not always the best judge of how useful or not his own inventions will be for those who use them."—Socrates, 420 BC

Did the Kindle "improve" upon the book? Do e-reading devices in general? Is the syndrome of eternal hardware upgrades an evolutionary given? Is the desire for speed, convenience, portability, and connectivity the natural and salutary "next step" for the reader always seeking the rich and rewarding experience of a great book? Socrates, for one, would say that something is not rationally so just because the CEO of Amazon claims it to be (we'll hear more from Socrates in the next chapter).

---

reading habits, although in ways more worrisome than a mere switch of delivery options.

nothing's_inevitable/

Since the launch of the Kindle, the reading public has been bombarded with media messages (and statements from Amazon) about the triumphant rise of e-books and e-reading. Overwhelmingly, such messages treat the displacement of print by digital text as an inevitability. In stark contrast to this doctrine of inevitability, however, are the findings of a study released in late 2012 by the Pew Research Center, which showed among American readers in all age groups a clear and persistent preference for ... print! The findings look like this:

- **Age 16-17:** 77% read a print book in the preceding year, while 12% read an e-book.
- **Age 18-24:** 78% read a print book, 21% an e-book.
- **Age 25-29:** 69% read a print book, 20% an e-book.
- **Age 30-39:** 79% read a print book, 25% an e-book.
- **Age 40-49:** 73% read a print book, 17% an e-book.

- **Age 50-64:** 71% read a print book, 15% an e-book.
- **Age 65+:** 64% read a print book, 8% an e-book.[9]

If this study is any indication, the reading of physical books still far outpaces digital books (or the use of e-reading devices like the Kindle). Compare this to the messages about e-books relentlessly conveyed to us through the media or from the digital gadget manufacturers, and in the resulting dissonance we find ideology at work.

"Nothing is inevitable," Marshall McLuhan once said, "so long as you are paying attention."

---

[9] See http://libraries.pewinternet.org/2012/10/23/younger-americans-reading-and-library-habits/

to_sum_up/

Ideology and technology always coexist and propel one another, even if technology's ideological impacts are unpredictable. As potential users of any new technology, we do best to interrogate the system of thought embedded in the tool or perpetuated by it, and to consider all the implications of such ideology.

//:read_me/

From *What Are People For?*
by Wendell Berry (North Point Press, 1990)

"The ongoing revolution of applied science known as 'technological progress'... has provided the means by which both the productive and consumptive capacities of people could be detached from household and community and made to serve other people's purely economic ends. It has provided as well a glamour of newness, ease, and affluence that made it seductive even to those who suffered most from it. In its more recent history especially, this revolution has been successful in putting unheard-of quantities of consumer goods and services within the reach of ordinary people. But the technical means of this popular 'affluence' has at the same time made possible the gathering of the real property and the real power of the country into fewer and fewer hands. ...

78

What is the purpose of this technological progress? What higher aim do we think it is serving? Surely the aim cannot be the integrity or happiness of our families, which we have made subordinate to the education system, the television industry, and the consumer economy. Surely it cannot be the integrity or health of our communities, which we esteem even less than we esteem our families. Surely it cannot be love of our country, for we are far more concerned about the desecration of the flag than we are about the desecration of our land. Surely it cannot be the love of God, which counts for at least as little in the daily order of business as the love of family, community, and country.

The higher aims of 'technological progress' are money and ease. And this exalted greed for money and ease is disguised and justified by an obscure, cultish faith in 'the future.' We do as we do, we say, 'for the sake of the future,' or 'to make a better future for our children.' How we can hope to make a good future by doing badly in the present, we do not say. ...

The question of the desirability of adopting any technological innovation is a question with two possible answers — not one, as has been commonly assumed. If one's motives are money, ease, and haste to arrive in a technologically determined future, then the answer is foregone, and there is, in fact, no question, and no thought. If one's motive is the love of family, community, country, and God, then one will have to think, and one may have to decide that the proposed innovation is undesirable. ...

I should ask, in the first place, whether or not I wish to purchase a solution to a problem that I do not have."*

---

* This passage is found in Berry's essay "Feminism, the Body, and the Machine"

key_questions/

1. Do you have any concerns about the ways digital books and e-reading devices are promoted or discussed in the media? If so, what are your concerns? If not, why not? How do these feelings influence your actions as a reader and buyer of books?

2. Consider McLuhan's observation, "The medium is the message," in the context of technology's relation to ideology. What might be some ideological "messages" of an e-reading device? How do these differ from the design-ideology of a printed book?

3. How do standards concerning what makes a good book differ from technological standards concerning what makes a good device?

4. Are e-reading devices the next phase of Gutenberg's revolution, or something entirely different?

further_recommended_reading/

- *The Image: A Guide to Pseudo-Events in America* by Daniel J. Boorstin
- *Media Unlimited: How the Torrent of Images and Sounds Overwhelms Our Life* by Todd Gitlin (Henry Holt, 2002)
- *The Influencing Machine* by Brooke Gladstone (W.W. Norton, 2011)
- *Living in Truth* by Václav Havel (Faber & Faber, 1986)
- *Brave New World Revisited* by Aldous Huxley
- *Privacy* by Garret Keizer (Picador, 2012)
- *The Net Delusion* by Evgeny Morozov (Public Affairs, 2012)

//:chapter_3/

Neuroplasticity

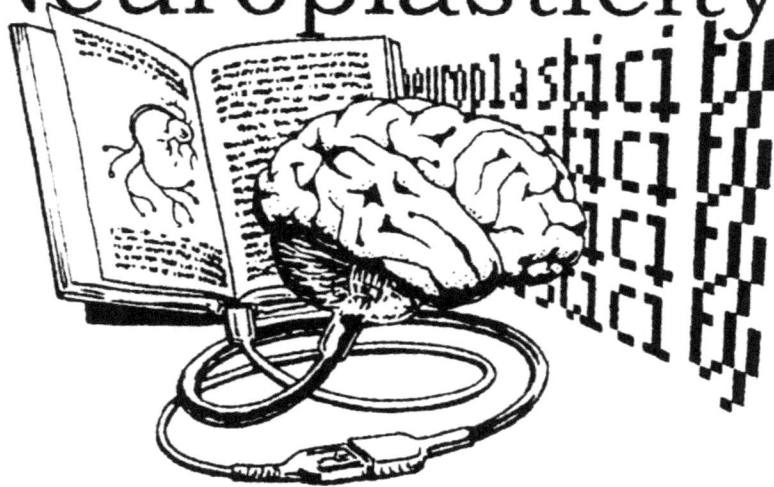

What Do a Story from Ancient Greece and Decades of Brain Research Have in Common?

a_story_about_consciousness/

This chapter concerns technology and the human brain. Here we move beyond ideology to the neurological impacts of our latest favorite tools.

Let's begin by returning to **Socrates**, our first guide for this chapter, and mentioned briefly in the previous one. The quote in response to Jeff Bezos comes to us from Plato's *Phaedrus,* which dates to about 375 BC and includes a dialogue Socrates once conducted concerning the "new" technology of writing in ancient Greece.[10] Socrates tells a story involving two characters: the ancient Egyptian god Theuth (said to be the inventor of writing) and the Egyptian King Thamus (Theuth's boss). One day, much in the style of Bezos pushing his Kindle, Theuth

---

[10] In fact, at the actual time of Socrates' dialogue, about 420 BC, the Greek alphabet had been around for roughly 300 years, but it was slow to catch on. The oral tradition which Socrates exemplified remained the central mode of communication and philosophical inquiry.

comes before Thamus to unveil his latest invention, proclaiming to his liege,

"Writing will make the Egyptians wiser and give them better memories. This invention will cure forgetfulness and folly!"

Thamus, who happens to enjoy his share of kingly wisdom, is a savvy consumer. He gives Theuth a skeptical, cautionary reply: "An inventor is not always the best judge of how useful or not his own inventions will be for those who use them." Thamus goes on to catalogue for his eager subordinate the possible undesirable effects of this new technology. Writing, he counsels Theuth, may be good for *reminding*, but not for *memory*. It will give people the false appearance of wisdom. They'll learn to hear without *absorbing* anything. They may seem to know everything while knowing nothing.

For those of us reading it today, the ironies in Socrates' story are immense. We know well the incredible usefulness of writing to humankind. Books have been described as breaking the bonds of human memory, and this is true. Moreover, we may recall that had Plato never taken the pains to record them in writing, the teachings of Socrates himself might have been lost altogether.

Amusing as all this is, can we say that Socrates is entirely off base in his arguments against writing? Wasn't he correct, for instance, in the prediction that literacy would diminish the powers of memory? The art of oratory as practiced in ancient Athens is long lost to us, after all. Today we have no traveling poets reciting the works of Homer in our public squares. And while some of us may retain, as great actors do, the cognitive capacity to memorize a leading part in a Shakespeare play, the *cultural emphasis* on the importance of memory has been lost. Nowadays, our books and tools remember *for* us.

For all of writing's indispensable benefits, still we can find ready examples affirming each of the arguments Thamus/Socrates makes against it:

- **Writing is good for *reminding*, but not for *memory*.** *Consider your grocery list. You arrive at the store only to realize you've left it at home; did you happen to memorize the items while writing them down?*

- **Writing will give people the false appearance of wisdom.** *Consider the teleprompter in the realm of politics.*
- **Writing, people will learn to hear without *absorbing* anything.** *Consider your class notes in school. As you listened to the instructor, did your notes do the remembering, or did you?*
- **Writing will make people seem to know everything while knowing nothing.** *Consider the bibliography at the back of any scholarly book.*

We needn't believe that we are better off without writing in order to see the validity of Socrates' arguments. And it is useful to grant that Socrates had a point because, in fact, he was making a larger observation about new technology in general. Two things in particular he recognized correctly: 1) that with the adoption of new tools, we change the nature of human consciousness, and 2) those who create the tools are not reliable predictors of their effects.

electrochemical_affinities/

Our second guide for this chapter is **Nicholas Carr**, author of the Pulitzer Prize finalist *The Shallows: What the Internet is Doing to Our Brains* (W.W. Norton, 2010). Engrossing and provocative, *The Shallows* looks seriously at how our latest technologies are changing the ways we think and behave by changing the actual physical structure of our brains.

"An intellectual technology asserts its influence by shifting the emphasis of our thought."
—Nicholas Carr

The last several decades of neurological research have given us a new understanding of how the brain works. Let's take a moment to consider what human brain function looks like.

Inside our brains we have approximately 100 billion nerve cells, or neurons. As we move through our day, these neurons are busily at work all the time, allowing us to perform each physical and mental function while also regulating emotions. In an active neuron, an electrical signal sparks the release of chemicals called neurotransmitters. Neurotransmitters carry electrical signals across the infinitesimal spaces between neurons (called synapses) and deliver the signals to the antennae-like dendrites of a nearby neuron, activating new signals in that neuron, or "overriding" earlier ones. As Carr explains, "It's through the flow of neurotransmitters across synapses that neurons communicate with one another, directing the transmission of electrical signals along complex cellular pathways. ...

Thoughts, memories, emotions — all emerge from the electro-chemical interactions of neurons."

A generation of study has clearly shown that frequent, repetitive behaviors strengthen the synaptic pathways in certain areas of the brain, leaving other less active pathways to "cool." In other words, our more habitual actions, thoughts, and emotions build up concentrated areas of specific cellular activity in our brains. To describe this process, neuroscientists employ a neat mnemonic aid:

Cells that fire together wire together.

It's a physical, chemical process whereby, you might say, we change our brains and our brains change us. Our behaviors influence the flow of neurological circuits (cells firing together). In turn, the electrochemical inclinations of those circuits increasingly lead us to seek certain behaviors over others (cells wiring together). All of which newly illuminates the old phrase

DENDRITES

NUCLEUS

AXON

AXON TERMINALS

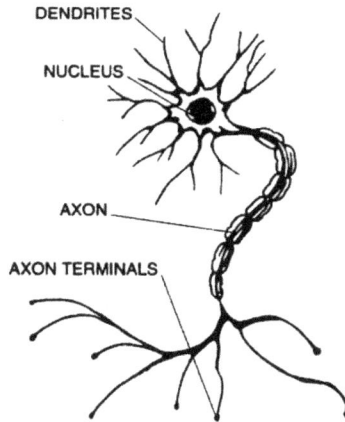

"ingrained behavior." Today, naturally, our ingrained behaviors include — often revolve around — our "interactive" technologies. The neurological effects of these tools and devices are unprecedented in the history of technological change, both because the devices have so quickly infused our daily lives, and because their use involves particular high-intensity, minute, repetitive behaviors (tapping keys, scrolling screens, visually scanning content) that promote rapid — and hard to reverse[11] — strengthening of certain cognitive functions over others. From a neurological standpoint, it is clear that these tools engender "new" functions (e.g., data entry reflexes) that differ substantially from the kind of "slow-burning" engagement or sustained attention that inherently characterize deep reading.

the_distracted_book/

"With the exception of alphabets and number systems," writes Nicholas Carr in *The Shallows*,

---

[11] Carr cites a great deal of research indicating that "plastic" in neuro-plasticity should not be confused with "elastic." Changes in brain function fostered by the "repetitive, intensive, interactive, addictive" stimuli of Internet technologies are not easily undone.

"the Net may be the single most powerful mind-altering technology that has ever come into general use. At the very least, it's the most powerful that has come along since the book. ... One thing is clear: if, knowing what we know today about the brain's plasticity, you were to set out to invent a medium that would rewire our mental circuits as quickly and thoroughly as possible, you would probably end up designing something that looks and works a lot like the Internet."

Granting the speed and power with which Internet technology is changing our brains, one might well contend that e-books are not the Internet, and that one may still read even long-form digital text without deleteriously modifying brain

97

function. E-reading, after all, is not the same as surfing the Web. Unfortunately, such a contention is less and less true. While e-reading devices were initially Web-browser-free and marketed as "dedicated" gadgets for long-form reading, today they are ever more frequently promoted as full capacity multimedia tablet computers, useful as much for checking e-mail, surfing the Web, or downloading movies and music as for digging into Tolstoy. Moreover, among younger readers of e-books, the most frequently used reading devices are not dedicated Nooks, Kindles, or Kobos. The 2012 Pew Research Center study mentioned in the last chapter includes in its findings the fact that among e-book readers age sixteen to twenty-nine, 55% utilized a personal computer for their reading device. Cell phones were the second most frequent means of e-book delivery, used by 41% of e-book readers in the same age group. Purpose-built e-reading devices themselves rated a distant third in usage, preferred by only 23%. Increasingly, e-books are read within the distraction-prone, multitasking environment of the Internet, thus the cognitive impact of digital reading promises to become all but indistinguishable from that of the Web; i.e., a profound qualitative shift in the reader's brain function.

a_prophetic_outcome/

Given the facts of neuroplasticity, we see that Marshall McLuhan, back in 1964, was more right than perhaps he could have known. "The medium is the message" indeed, in that a medium is a *neurological message* capable of changing the structure of our brains and thus the physiological context of our thought.

Cells that fire together wire together.

⇧ ⇧ ⇧

The Medium is the Message.

"We shape our tools," McLuhan said, "and thereafter our tools shape us."

99

qualities_of_experience/

How might we specify, characterize, and define the ways in which the new medium of e-text is qualitatively changing our thought processes and reading methods? The full implications remain to be seen. As frequent users of screen technology, however, we each may weigh for ourselves the felt effects of prolonged exposure to our devices. If honest, we might admit that we're more, well, *jittery.* Perhaps, too, we are generally less adept at deep, sustained, active attention than we used to be. Carr offers a powerfully broad observation:

> "For the last five centuries [i.e., since Gutenberg] ... the linear, literary mind has been at the center of art, science, and society. ... It's been the imaginative mind of the Renaissance, the rational mind of the

Enlightenment, the inventive mind of the Industrial Revolution, even the subversive mind of Modernism. It may soon be yesterday's mind."

to_sum_up:_a_plate_is_not_a_plate/

Though a work by Thomas Hardy may be identical in its wording on both page and screen, the medium of delivery inevitably changes the way we read and absorb the work. E-reading is not simply a matter of "same food, different plate," as many techno-enthusiasts would suggest, for here the "plate" subtly but significantly changes the properties of the "food" and thus alters the food's absorption into one's physiology. Our brains process what is virtual and manipulable quite differently from what is printed and fixed. Moreover, the most common e-reading environment is the interruption-laden Internet, known to deplete our cognitive capacities for deep sustained attention.

Given a choice between text on a page in a printed book or text transmogrified to a pixelated image on an electrically illumined screen, a choice between text at rest in the "low-tech" medium of bound paper or text rendered fluid in the high-intensity environment of a gadget with buttons, we should very seriously consider the significant distinction in the behaviors required of us as readers in each case, and the respective results of these behaviors.

//:read_me/

From *The Shallows: What the Internet is Doing to Our Brains*
by Nicholas Carr (W.W. Norton, 2010)

"When the Net absorbs a medium, it re-creates that medium in its own image. ... It injects the medium's content with hyper-links, breaks up the content into searchable chunks, and surrounds the content with the content of all the other media it has absorbed. All these changes in the form of the content also change the way we use, experience, and even understand the content.

Scrolling or clicking through a Web document involves physical actions and sensory stimuli very different from those involved in holding and turning the pages of a book or magazine. Research has shown that the cognitive act of reading draws not just on our sense of sight but also on our sense of touch. It's tactile as well as visual. 'All reading,' writes Anne

Mangen, a Norwegian literary studies professor, is 'multi-sensory.' There's a 'crucial link' between 'the sensory-motor experience of the materiality' of a written work and the 'cognitive processing of the text content.' The shift from paper to screen doesn't just change the way we navigate a piece of writing. It also influences the degree of attention we devote to it and the depth of our immersion in it.

... When a printed book — whether a recently published scholarly history or a two-hundred-year-old Victorian novel — is transferred to an electronic device connected to the Internet, it turns into something very like a Web site. Its words become wrapped in all the distractions of the networked computer. Its links and other digital enhancements propel the reader hither and yon. ... The linearity of the printed book is shattered, along with the calm attentiveness it encourages in the reader.

... The Net delivers precisely the kind of sensory and cognitive stimuli — repetitive intensive, interactive, addictive —

that have been shown to result in strong and rapid alterations in brain circuits and functions.

... As the time we spend hopping across links crowds out the time we devote to quiet reflection and contemplation, the [neural] circuits that support those old intellectual functions and pursuits weaken and begin to break apart. The brain recycles the disused neurons and synapses for other, more pressing work. We gain new skills and perspectives but lose old ones."

key_questions/

1. Neurologically, our responses to printed text are different than our responses to digital text. In what other ways, personal or social, is a print book's physical nature important?

2. As we continue our intensive use of interactive screen technologies, how might our resultant neurological changes affect **a)** the way we write, **b)** our public discourse, and **c)** our civic and cultural life?

3. Considering the findings of neuroscientists as cited by Nicholas Carr and others, do you personally feel an urgency to reduce or otherwise mitigate your own use of interactive screens? If so, what steps might you take toward that end? If not, why not?

further_recommended_reading/

- *The Brain that Changes Itself* by Norman Doidge (Penguin, 2007)
- *iDisorder* by Larry Rosen (Palgrave Macmillan, 2012)
- *Alone Together* by Sherry Turkel (Basic Books, 2011)

//:appendix/

some_more_key_questions_about_e-books/

- In 2009 Amazon.com removed, without warning or consent, digital copies of George Orwell's *1984* from Kindle users' e-libraries. In 2010 Apple refused to publish as an e-book a graphic novel adaptation of James Joyce's *Ulysses* containing drawings of human nudes. Both companies, faced with public pressure, reversed their actions. What concerns do such incidents raise for you personally?

- In what ways is digital text vulnerable to manipulation, ad hoc revision, etc., and what might this mean for the integrity of our information systems or literary heritage?

- What will e-books ultimately mean for bookstores and

libraries? Is it a role of libraries to protect print and champion its continuance?

- What would a digital age "Renaissance" look like?

- Author Milan Kundera has said: "When everyone wakes up as a writer, the age of universal deafness and incomprehension will have arrived." What might be made of this comment, in a world of digital texts?

- Would a mass changeover to e-books mean an unprecedented consolidation of media? Why or why not?

- All e-books purchased through Amazon, Apple, Barnes & Noble, etc., are subject to terms of service explicitly stating that the digital texts remain the property of the corporation and are subject to "modification" or "discontinuation" at any time. Readers do not *own* e-books, but lease them under these companies' "end-user license agreements," (or EULAs). What long-term questions does this raise for you?

- Does large-scale digitization and online reading significantly endanger privacy? If so, can legislation address this?

- How might a predominantly electronic publishing culture affect — for good or ill — those in poverty?

- Is an electronic publishing model truly more environmentally sound than traditional publishing? (See http://www.danielgoleman.info/e-reader-versus-book-the-eco-math/ and http://www.themillions.com/2012/05/are-ereaders-really-green.html)

- What do patterns in past technological developments suggest about the ways technologies and traditions tend to displace or complement one another (orality to literacy; radio to TV; theater to film)?

- How have the reading habits and modes of discourse of the general population changed since the Victorian era (or perhaps just since the Age of Criticism) and what part do the Internet and e-reading devices seem to be playing in that history?

- What constitutes a book? What constitutes one's experience of a book? What constitutes a conversation about a book? What constitutes authorship?

- Under what conditions are authors entitled to/deserving of a livelihood?

- Is literature, in its gestation and digestion, fundamentally interior/intuitive/emotional, or social/rational? Which way do e-books and e-reading lean?

- Can the technology of the printed book be improved, and do e-readers like the Kindle represent such an improvement or something else altogether? Do we do injustice to both e-reading devices *and* books to compare them under these terms?

- Does digitization equal a destruction of books? Or could it lead to the unthinking destruction/disposal of books? How so? How not so?

- How does the removal of physical texts through digitization differ from the systematic, censorship-motivated disappearance of books in despotic states? How significant are the distinctions?

113

Add your own questions or concerns about e-books here:

more_for_reflection_&_discussion/

"Containers and Their Contents" : a debate between Clay Shirky (*Cognitive Surplus*) and Nicholas Carr (*The Shallows*) about which old media forms will survive in the new digital age.   http://www.roughtype.com/?p=2315

"E-Books That Read You" : "If everything gets informed by data, do we lose something fundamental about literature?" "Does the availability of [e-readers]...put us all in danger of being in suspicion?"   http://www.onthemedia.org/2012/jul/13/ebooks-read-you/

Speech by John Updike about the printed book, Book Expo America 2006 : http://bookexpocast.com/2006/05/26/bea-2-john-updike-speech/

www.ingramcontent.com/pod-product-compliance
Lightning Source LLC
Chambersburg PA
CBHW081657270326
41933CB00017B/3200